Fire Department Response to Biological Threat at B'nai B'rith Headquarters Washington, DC

This is Report 114 of the Major Fires Investigation Project conducted by Varley-Campbell and Associates, Inc./TriData Corporation under contract EMW-94-C-4423 to the United States Fire Administration, Federal Emergency Management Agency.

FEMA

Department of Homeland Security
United States Fire Administration
National Fire Data Center

U.S. Fire Administration Fire Investigations Program

The U.S. Fire Administration develops reports on selected major fires throughout the country. The fires usually involve multiple deaths or a large loss of property. But the primary criterion for deciding to do a report is whether it will result in significant "lessons learned." In some cases these lessons bring to light new knowledge about fire--the effect of building construction or contents, human behavior in fire, etc. In other cases, the lessons are not new but are serious enough to highlight once again, with yet another fire tragedy report. In some cases, special reports are developed to discuss events, drills, or new technologies which are of interest to the fire service.

The reports are sent to fire magazines and are distributed at National and Regional fire meetings. The International Association of Fire Chiefs assists the USFA in disseminating the findings throughout the fire service. On a continuing basis the reports are available on request from the USFA; announcements of their availability are published widely in fire journals and newsletters.

This body of work provides detailed information on the nature of the fire problem for policymakers who must decide on allocations of resources between fire and other pressing problems, and within the fire service to improve codes and code enforcement, training, public fire education, building technology, and other related areas.

The Fire Administration, which has no regulatory authority, sends an experienced fire investigator into a community after a major incident only after having conferred with the local fire authorities to insure that the assistance and presence of the USFA would be supportive and would in no way interfere with any review of the incident they are themselves conducting. The intent is not to arrive during the event or even immediately after, but rather after the dust settles, so that a complete and objective review of all the important aspects of the incident can be made. Local authorities review the USFA's report while it is in draft. The USFA investigator or team is available to local authorities should they wish to request technical assistance for their own investigation.

For additional copies of this report write to the U.S. Fire Administration, 16825 South Seton Avenue, Emmitsburg, Maryland 21727. The report is available on the Administration's Web site at http://www.usfa.dhs.gov/

U.S. Fire Administration
Mission Statement

As an entity of the Department of Homeland Security, the mission of the USFA is to reduce life and economic losses due to fire and related emergencies, through leadership, advocacy, coordination, and support. We serve the Nation independently, in coordination with other Federal agencies, and in partnership with fire protection and emergency service communities. With a commitment to excellence, we provide public education, training, technology, and data initiatives.

 FEMA

ACKNOWLEDGEMENTS

The United States Fire Administration Major Fire Investigation Team would like to thank the many members of the District of Columbia Fire and Emergency Medical Services (DCFEMS) Department who provided information for this report. We are grateful to the following persons for sharing their experience:

DCFEMS

    Chief Donald Edwards

    Assistant Chief Floyd Madison

    Deputy Chief Joseph Herr, Training Division

    Battalion Chief Thomas Johnson, Special Operations

    Battalion Chief William Mould, Battalion 6

    Hazmat Unit and Engine 12, including

        Battalion Chief Tom Herlihy

        Lieutenant Anthony Culver

        Firefighters Shelley Nickelson, Jerry Barnes, Rob Small, Tim Clark

    Lieutenant Keith Harter, Jr., Communications Division

DC Office of Emergency Preparedness

    Dr. Michele Penick, Chief of Plans and Training Division

Washington Metro Council of Governments Metropolitan Medical Strike Team

    Captain Mike Moultrie, Arlington County (VA) Fire Department

Federal Bureau of Investigation

    Debbie Stafford, Unit Chief, FBI Headquarters

    James Rice, Supervisory Special Agent, Washington Field Office/Joint Terrorism Task Force

Centers for Disease Control

    Agency for Toxic Substances and Disease Registry

    Dr. Richard Knudsen, Biosafety Branch

Department of Health and Human Services

    Dr. Robert Knouss, M.D., USPHS, Director, Office of Emergency Preparedness

TABLE OF CONTENTS

Fire Department Response to Biological Threat at B'nai B'rith Headquarters
Washington, D.C.
April 24, 1997

Investigated by: Jeff Stern

Local Contacts: Fire Chief Donald Edwards
District of Columbia Fire and EMS Department
1923 Vermont Avenue, N.W.
Washington, D.C. 20001

Deputy Chief Joseph Herr, Training Officer
DCFEMS Training Division
4600 Shepherd Parkway, S.W.
Washington, D.C. 20032

Battalion Fire Chief William Mould
6th Battalion
1018 13th Street, N.W.
Washington, D.C. 20001

Battalion Fire Chief Thomas Johnson, Special Operations
DCFEMS Training Division
4600 Shepherd Parkway, S.W.
Washington, D.C. 20032

OVERVIEW

On April 24, 1997, the District of Columbia Fire/EMS Department (DCFEMS) responded to the Washington, D.C. offices of B'nai B'rith, an international Jewish organization, for a suspicious package. The package had been mailed to B'nai B'rith and contained a Petri dish labeled with wording which led local emergency responders to suspect the package might contain Anthrax and Yersinia, both disease-causing bacteria that have been used as biological weapons. Several occupants of the building complained of dizziness and headaches, additional factors that indicated the possible presence of a chemical agent as well.

1

The DCFEMS established a perimeter around the site where the package was located, attempted to protect occupants of the B'nai B'rith building by isolating them in place, established a command post, and set up hazardous material decontamination procedures. After consultation with numerous national agencies, including the Centers for Disease Control (CDC) Biosafety Branch, responders engaged in Hazmat operations to secure the package and its contents, which were then sent to a Federal laboratory in Bethesda, Maryland. Laboratory tests revealed that the contents of the package were non-hazardous.

During the operation, DCFEMS decontaminated approximately 30 people for exposure, including civilians, police officers, and fire-rescue personnel. One security guard suffered chest pains during the incident and was transported to a local hospital for a heart attack following decontamination.

Though there have been several chemical and biological incidents over the past few years, this incident received National media exposure and was broadcast life on CNN and other news networks, generating discussions among emergency responders as to the actions that took place. Though the threat was a hoax, the incident revealed many lessons for the fire service to share in preparation for any future threats.

SUMMARY OF KEY ISSUES

Issue	Comment
Incident command	Although an incident command system (ICS) was established, weaknesses in operating a unified command post among all the responding agencies led to some uncoordinated actions. Some agencies, such as the health department, had not received prior ICS and emergency management training. The incident confirmed the advantages of conducting emergency management training among diverse city agencies.
Communications with victims	It is essential to establish a single source of communications between victims of an incident and the emergency responders.
Training of first responders	Police, Fire and EMS personnel should be trained to recognize and assess NBC hazards. Their initial actions will affect the course of the incident. Courses that cover recognition and response to these incidents should be provided to all potential first responders.
Biological agents differ from chemical agents	Departments must develop protocols for handling each. Different action plans may be necessary depending upon both the type and the form of the hazard, biological or chemical, airborne, liquid or solid.
Resources	Incidents involving potential chemical or biological agents will be resource intensive. They will require the assistance of multiple agencies, many personnel, and specialized equipment. Departments should know where these resources are how to get help quickly.
Access to emergency information	Delays occurred as the department attempted to access information on the potential hazardous substances. Since this incident, the Federal government has established the National Response Center (1-800-424-8802) as a hotline number for incidents involving chemical, biological, or radiological agents.
Public information	Strong public information and media relations is essential to get the corrected information disseminated to the public in a timely fashion. Regular briefings should be held, media access should ba appropriately limited to avoid interfering with operations, and an articulate technical advisor should be assigned to assist with media inquiries.

DCFEMS PROFILE

The District of Columbia Fire and Emergency Medical Services Department provides services to the 600,000 citizens of Washington, D.C., including the government of the United States and millions of visitors annually. The DCFEMS service area includes numerous Federal facilities, including the White House, the Capitol, and the Supreme Court. Services are provided by over 1,400 personnel from 32 fire stations covering 63 square miles. Mutual aid agreements exist with surrounding jurisdictions.

INCIDENT REVIEW

On April 24, 1997 at approximately 11:00 a.m. a person working for the Washington, D.C. office of B'nai B'rith discovered a suspicious looking package that was leaking a reddish fluid in the mailroom. The director of security for B'nai B'rith, a retired police officer, opened the package and found a Petri dish and a threatening note. He took the package and its contents outside the B'nai B'rith building and placed it on a small grass lawn. He then dialed 9-1-1 and requested police.

The DC Metropolitan Police Department (MPD) responded to the incident. Police officers and the MPD Bomb Squad were on the scene for about one hour. Assistance from the fire department was requested when the police determined that the package contained a Petri dish labeled *Anthracis Yersinia*, a misspelled combination of the words Anthrax and Yersinia. Both are disease-causing bacteria that can be used in the production of biological weapons (see Appendix C).

At 12:20 p.m. the DCFEMS Communications Center dispatched Engine 16, Truck 9, the Hazmat Unit, and Battalion Fire Chief 6 on a local alarm for the report of an unknown package. The MPD personnel, including on member of the Joint Terrorism Task Force for the Washington area, reported to the fire department that the package might contain Anthrax. Battalion 6 instructed E-16 and T-9 to stage at 15th Street and Rhode Island Avenue while he and the Hazmat Unit responded to Scott Circle, about a block away from the location of the package in front of the B'nai B'rith Building. At 12:46 p.m., he requested the special response of the Hazmat team captain, who was serving as acting Battalion Chief 1 for the shift.

Some people in the building complained of headaches, dizziness, and other minor ailments. The rapid onset of these symptoms led responders to believe that the package might also contain some kind of chemical agent. An action plan was developed that treated the incident as a possible chemical hazard, as well as a biological one.

Initial efforts were made to contact Chemtrec (the Chemical Manufactures' Association emergency information resource), the United States Army Medical Research Institute For Infectious Disease (USAMRIID) at Ft. Detrick in Maryland, and the Centers for Disease Control (CDC) in Atlanta, Georgia. Chemtrec had no information to pass on to the fire department, and the responders had to wait for return calls from USAMRIID and CDC. Battalion 6 requested a Hazmat Task Force, the full DCFEMS hazardous materials box assignment, bringing Engine 12, Truck 4, Rescue Squad 1, Medic 17 and EMS 3 (an EMS supervisor) to the scene. A hot zone surrounding the area was established.

Once the Hazmat Task Force units were on the scene, E-16 and T-9 were returned to service. Other agencies responded, including the Federal Bureau of Investigation (FBI). Under Presidential Decision Directive 39 (PDD-39), which gives jurisdictional authority for "crisis management" to the FBI in terrorist incidents, the FBI assumed overall command of the incident.[1] This was done in a quiet and

---

[1] The Federal Emergency Management Agency (FEMA) is given jurisdiction over "consequence management" in the same directive.

unassuming manner. The FBI Supervisory Special Agent in charge of the Joint Terrorism Task Force was present at the incident command post and offered assistance, as they simultaneously began a criminal investigation into the incident. Battalion 6 retained the key operational role in charge of mitigating the incident.

After some discussion over options, the fire department decided upon an action plan to shelter in place 109 civilians in the B'nai B'rith building. The fire department further isolated the scene, expanding the hot zone to include a one block area surrounding the B'nai B'rith building. Occupants of a hotel across the street from the B'nai B'rith building were also sheltered in place, and the HVAC systems for both buildings were shut down. At this time, approximately 1:05 p.m., the Metro/Safety Battalion Chief arrived on the scene, as did the DCFEMS Mass Casualty Unit. Additional responders included the deputy fire chief and the fire chief.

The CDC eventually provided information to the Hazmat team, advising them in protective measures and decontamination procedures. Based on the initial information, CDC surmised that the package might not pose a great threat because the form of the suspected contents (in a Petri dish) would not enable airborne pathogens to be transmitted immediately, if at all. However, since some occupants of the building exhibited various symptoms, Hazmat personnel decided to err on the side of caution, suspecting that a chemical agent might also be present.

The fire department Hazmat sector officer raised the protective level used by Hazmat personnel. Level A encapsulated entry suits were worn by entry personnel, and Level B suits were worn by the decontamination team. The CDC advised that decontamination with a 1.0% bleach solution would be effective.

The Hazmat Task Force was briefed and personnel were assigned to sectors. Battalion 1 adjusted the DCFEMS Hazmat procedures after conferring with the CDC. The lieutenant from Truck 4 was assigned to Safety Engine 12's crew and was assigned to the Decon Sector, with the lieutenant from Rescue Squad 1 assigned as the Decon Officer. EMS 3 was assigned to handle the quarantine of civilians and police officers in the hot zone, and Medic 17's paramedics were assigned to conduct post decontamination medical evaluations.

Additional resources arrived on the scene, including the DC Office of Emergency Preparedness (DCOEP), the DC Department of Health, and a liaison from the United States Secret Service Technical Security Division. Two Hazmat personnel were then dressed as an entry team, with two additional personnel dressed as a back-up team. The entry team, under directions from Federal experts on the scene, secured the package in double bagging and placed it in a five-gallon packing drum. The entry crew and the secured package were then decontaminated. The decontaminated package was turned over to the FBI, which transported it to the National Naval Medical Center in nearby Bethesda, Maryland, for laboratory analysis.

Personnel then waited for the results of the testing. During this time, a security guard in the quarantine area developed chest pains. He was carried on a chair through the decontamination corridor and then transported to a local hospital for treatment at 3:25 p.m. Also during this waiting period, several MPD officers became upset with instruction that they undergo decontamination. The officers had become aware that the media was broadcasting live pictures from cameras positioned on top of a nearby building. The officers refused to disrobe and undergo decontamination. One of the officers struck the EMS Lieutenant assigned to the quarantine area. High-ranking police officials were asked to help get the officers to comply with the procedures and, eventually, the officers were decontaminated. Firefighters were sent to check on the welfare of the civilians who had been sheltered in place.

The DCFEMS Rehab Unit was brought in to provide a waiting area for decontaminated civilians and police officers. Kosher food was ordered and delivered for the civilians still isolated in the B'nai B'rith building. At approximately 8:00 p.m., almost nine hours after the package was first discovered, results of the analysis of the package revealed that no chemical or biological threat was present. A final press conference was held, the civilians being protected in the building were released. The fire department terminated operations at 9:00 p.m.

Responders:

Agencies responding to or assisting at the B'nai B'rith incident included:

Local:

Fire/EMS Department

Metropolitan Police Department

Department of Consumer and Regulatory Affairs

Department of Health

Office of Emergency Preparedness

Office of Communications

Department of Public Works, Water and Sewer

Office of the Mayor

Federal:

Federal Bureau of Investigation

United States Secret Service

Public Health Service Agency for Toxic Substances Disease Registry

Centers for Disease Control Biosafety Branch

National Naval Medical Center

United States Park Police

Agencies which were notified or placed on standby:

US Marine Corps Chemical/Biological Incident Response Force

American Red Cross

DC Housing Authority

DC National Guard

Metro Transit Authority (bus and rail)

Metro Traffic Control

DC Public Schools

Military District of Washington

Federal Emergency Management Agency

Department of Human Services

Users of Emergency Preparedness Network/NAWAS

ANALYSIS

DCF/EMS Preparedness and Response

Few fire departments are prepared to handle the scope of a potential terrorist chemical/biological attack. The DCFEMS was able to utilize existing response plans and adapt their own Hazmat policies and procedures to deal with the scope of this incident. The fire department incident commanders made appropriate decisions based upon their knowledge of the perceived threat, and consistent with their training and experience. First responding fire units were staged at appropriate distances while the battalion chief and the Hazmat Unit sized up the situation. A hot zone was established quickly. Unfortunately, these efforts were compromised by a few MPD officers on the scene who did not understand the serious nature of the potential chemical/biological threat and treated the incident more as a bomb scare. These officers initially crossed in and out of the hot zone, and could have contaminated themselves and other personnel had this been a real incident.

DCFEMS personnel performed admirably given the limited amount of information provided regarding what might be contained in the threat package. The Hazmat task force units provided trained hazardous materials technicians and specialists to conduct mitigation operations. Initial responding units (Engine 16 and Truck 9) that did not have trained Hazmat personnel were placed in service as quickly as possible to cover responses to other areas of the city.

The department was able to effectively utilize general guidelines it had established only a few months prior to this incident (Appendix A).

Hazmat Operations

The Hazmat operations were conducted in accordance with Nationally acceptable practices (OSHA 1910.120) for a fire department operating on the scene of an incident. It should be noted that, afterwards, some scientists and physicians expressed concerns as to how the department handled the contaminated package. These expressions ranged from comments that there was no real risk and the department's actions were overkill, to complaints that the process was not thorough enough. The fact is that the fire service personnel acted conservatively, after receiving diverse and contradictory information from many different sources. Personnel were presented with a leaking package containing an unknown substance, civilians who claimed to be symptomatic, and a threatening letter from an unknown perpetrator. The department operated after consulting the CDC, and (due to the perceived threat of a possible *chemical or biological* agent) increased the level of protection used by personnel. Hot, warm, and cold zones were established, and a quarantine area was set up for contaminated civilians and first responders.

Limitations in equipment did hamper operations. The lack of tents for the discreet decontamination of civilians forced several people to disrobe in front of television cameras. However, decontamination procedures have traditionally been targeted for the decontamination of fire service personnel and victims under emergency conditions. The department should not be faulted for the live broadcasts by local and National media which were beyond the fire department's ability to control. Many departments are, as a result, now taking the issue of the public's modesty into consideration in developing action plans for decontamination operations.

There was also a problem with an inadequate number of Hazmat personnel. The limited number of trained Hazmat technicians forced many of the personnel tasked with decontamination and entry operations to remain suited up or to rotate through assignments over a long period of time.

Communications were also difficult on several levels. Hazmat entry teams were forced to use hand signals because their suit radios did not function effectively. The team did not have enough cell phones, fax machines, or a computer to access information in a rapid and systematic fashion. The DCFEMS has made several steps to address these shortfalls (see Initiatives, below).

Protect-in-place or evacuate to a safe haven? – The review of this incident revealed disagreements between fire service responders and health service professionals at both the local and National level as to whether the occupants of the B'nai B'rith building should have been sheltered in place. Traditional fire service and Hazmat actions often lead to the isolation of an incident, including the protection in place of civilians. The DCFEMS followed this long established procedure in developing their action plan. Both the civilians in B'nai B'rith and in the hotel across the street were isolated and the HVAC systems were shut down.

Some health service providers disagree with this tactic for a biological incident. The contention among some health service experts is that people must be evacuated from a possibly contaminated area (in this case, the B'nai B'rith building) and moved to a safe haven where they can either be quarantined or treated prophylactically with medications or antibiotics. By isolating people in an unventilated and possibly contaminated area, the victims are in effect exposed to any toxic biological pathogen for an extended duration. The perception is that the people isolated in place have been effectively written-off by emergency response personnel. Some respected experts have even suggested that, because there is no true decontamination process for exposure to a biological agent, possible victims should be simply treated, observed, or sent home. The view is that – unlike chemical agents – unless a biological agent is present in an aerosolized form, transmission of the agent from person to person is less likely.

Further discussion and research is required on this topic. Ultimately, such decisions will be at the discretion of the incident commander, after conducting a thorough risk assessment based the unique nature each incident, to develop and decide upon the best course of action.

EMS Operations

Initial reports indicated that several people were complaining of dizziness and headaches. Efforts were made to isolate and assess all people who might have been exposed. After the decon area was set up, these people were decontaminated and either held in a bus or in a quarantine area. The George Washington University Medical Center and the Washington Hospital Center were alerted of the situation and prepared to receive patients. Several personnel were evaluated and released from these hospitals; one patient was admitted to a hospital after suffering chest pains while waiting in the hot zone. More thorough interaction with the civilians protected-in-place in the B'nai B'rith building, (including possible prophylactic treatment with medications), might be a consideration in any future incident of a similar nature.

Police Operations

The MPD performed their police functions well, isolating the scene and limiting access. However, there were problems associated with underestimating the magnitude of the chemical/biological threat posed by the incident. The police bomb squad did not notify the fire department until over one hour into the incident. Some police personnel did not cooperate with the directives of the fire and EMS personnel on the scene, making isolation of the hot zone difficult in the early stages of the incident. The officers could have contaminated themselves, police, and EMS crews had the threat materialized. MPD personnel had received no training for Hazmat or chem/bio response.

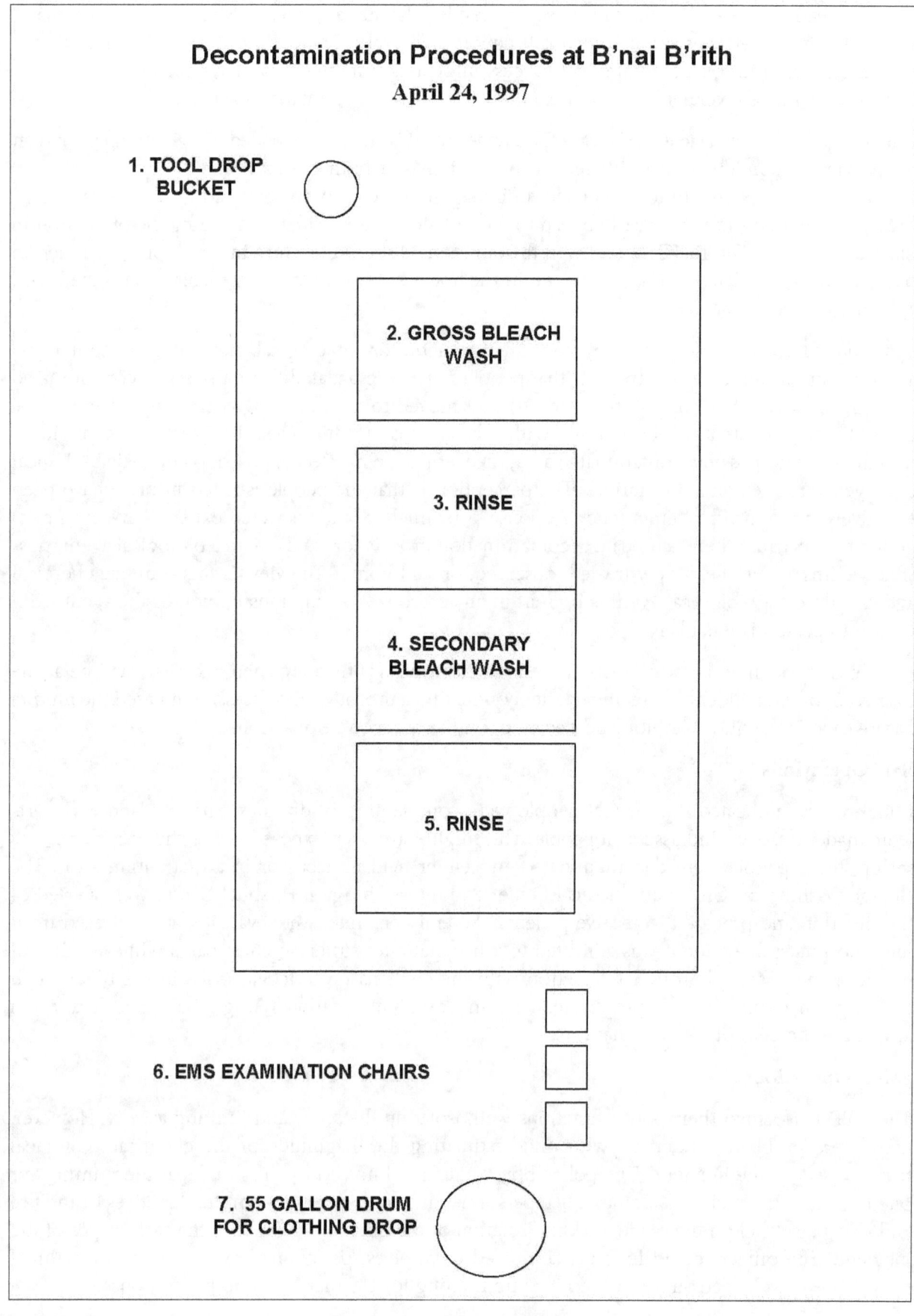

Interagency Coordination and Command Post Operations

Interagency coordination and unified command post operations were problematic during this incident. Key operational personnel (Fire/EMS, FBI, MPD, and Secret Service) had the benefit of previously working together during major events (such as the Presidential Inauguration), and were familiar with operating in an incident command system. Other agencies such as the DC Health Department, however, did not share that previous experience and did not have formal ICS training in unified command and emergency operations. The result was a fragmented operation in an overcrowded command post. While DCFEMS and FBI personnel were developing action plans, health department personnel were calling people in the B'nai B'rith building on their own, giving separate advice, and attempting to handle the situation as a public health operation. At on point, health officials told the people inside the building that they could leave, contrary to the instructions of the fire department incident managers. The local health officials, in turn, were operating under instructions they were receiving from their own National technical experts. The separate efforts led to different strategic decisions and action plans, and dramatically illustrates why it is critical for all responding agencies to operate under a unified approach during such incidents.

Public Information (PIO) Functions

The fire and police department's public information officers responded to handle the crush of media. An initial media area was established across the street from B'nai B'rith and later moved further away from the hot zone. Two media briefings were held during the event. More regular media updates, especially given the ongoing "live" broadcast of the incident, could have been helpful. Some people in B'nai B'rith were notified of the unfolding incident via associates who saw the incident coverage on television and called their colleagues in the building while the event was unfolding.

A close relationship with the media is helpful during almost any incident, but especially during chem/bio incidents. During this incident, the live broadcasts inhibited the ability of Hazmat crews to decontaminate people whose concerns about personal modesty appeared to outweigh their perception of life-threatening risk. In a real threat, the media can provide critical assistance by alerting civilians to necessary emergency actions. The department PIO can encourage media representatives to be sensitive to the victims' privacy concerns, explain technical operations as they are occurring, and relay important information to the public through the media.

Metropolitan Strike Team

The Washington Metropolitan Area Council of Government's Metropolitan Medical Strike Team was established in 1997 under a US Public Health Service grant to assist first responding agencies with the mass decontamination and treatment of victims of weapons of mass destruction. The Washington area MMST consists of police officers, fire department Hazmat personnel, and paramedics from the District of Columbia, Arlington County (VA), Fairfax County (VA), Alexandria (VA), Montgomery County (MD), and Prince George's County (MD). While members of the MMST had received training at the time of the B'nai B'rith incident, the strike team was not yet operational. Thus, it was not a viable resource for DCFEMS, though some personnel and equipment could have been utilized had they been requested by the DC responders. The DCFEMS has incorporated plans for activating MMST resources into its plans, now the team is fully operational.

It is important to recognize that the function of MMST is as a back-up to the first responding agency. The local fire, EMS, and police will be the first on the scene, and their initial actions will affect the overall magnitude and scope of the incident.

LESSONS LEARNED

Many valuable new lessons have been learned from this incident. Others have been documented before, but are worth reiterating.

1. **Perhaps the most important lesson is that a true chemical/biological incident will quickly overwhelm even the most prepared responders; it is essential to call early for additional EMS and Hazmat resources, including mutual aid assets.**

    Had this incident been a true terrorist attack, producing a substantial number of victims, DC would have needed assistance from surrounding Hazmat teams and EMS systems available through existing mutual aid agreements. Quick response of a variety of technical Federal resources was available because the incident occurred in the Nation's capital. Had the incident occurred elsewhere, it is likely that Federal resources would have taken much longer to arrive.

2. **A supply of vaccines and antibiotics should be available in stockpiles for responders and victims from a chemical or biological attack.**

    Stockpiles of nerve agent antidotes for chemical attacks, and medications for victims exposed to biological agents are only now being established. Vaccinations against some biological threats such as Anthrax are expensive and of yet unquantified value, though the military is inoculating many of its personnel. Treatment is available for many victims of bacteriological agents through prophylactic administration of antibiotics immediately after exposure. Limited treatments are available for those exposed to viral agents. Currently, it can take substantial time to collect and distribute the available medications and antidotes. Therefore, the identification of such resources should be an integral part of response plans.

3. **Senior fire and EMS officers need real time information on evolving incidents which include chemical and biological threats.**

    Fire department leaders are often kept "out of the loop" of intelligence reported to Federal law enforcement agencies or local police regarding valid threats of violence. In this incident, information gathered at the scene regarding the nature of the incident, the potential biological agent, and other police information was passed on to fire/EMS commanders in a timely fashion after the unified command was established. Prior to that, however, police units were on the scene for over one hour without informing the fire department of the incident.

4. **This incident demonstrates the importance of multi-agency cooperation under the unified command structure during major incidents.**

    At this incident, the agencies such as fire/EMS and FBI that had previously worked together in a command structure were able to operate in an efficient and unified means to achieve common goals. The agencies that had not previously worked with the fire department, such as the D.C. Health Department, had different information and approaches to the incident, and gave conflicting information to the people in the B'nai B'rith building. Unified command and exchange of key information and technical resources are critical factors for successful response.

5. **All first responders, including police, should be trained to size-up a potential chemical/biological incident and call for appropriate resources.**

    The lack of knowledge among MPD responders about the dangers posed by this threat could have led to unnecessary deaths had the threat been real. The delay in calling for fire department resources after the initial police response could have further endangered innocent civilians and severely inhibited the fire department's ability to contain the incident. All police, fire and EMS personnel should receive first response training in recognition and assessment for these threats. In addition, it is important that responders be trained to recognize the different containment actions required for various chemical or biological threats. Airborne agents, for example, should be approached differently than other forms, such as solid agents. The training of responders in recognizing these differences is essential to accomplishing an appropriate and effective intervention.

6. **Security guards and civilians in high-risk buildings and areas should be trained in the proper course of action for a suspicious package or device. The fire department or other local emergency responders are the logical people to conduct this training.**

B'nai B'rith is a clear target hazard for a terrorist incident (and actually was the site of a hostage incident in the 1970's). If the security officer had known to leave the suspicious package in place in the mailroom, and then call 9-1-1, this incident may have been contained to an enclosed area, greatly facilitating exposure control, containment, and mitigation operations.

7. **Biological agents present differently than chemical agents. Public health departments response plans should include the capability to recognize the differences and address biological and chemical attacks.**

While this incident was properly handled as a local chemical/Hazmat situation, a biological weapon release would likely sicken and kill victims over the course of several days. Anthrax, for example, mimics flu symptoms and has a 48-72 hour incubation period. A biological attack would reveal itself in the form of many people coming into emergency departments or being transported to hospitals with flu-like symptoms, or worse. A chemical agent will likely cause immediate and localized casualties. Mechanisms should be developed and put in place for public health departments and EMS services to identify when they are dealing with exposures that could be related to a biological attack, as opposed to a chemical attack.

8. **While Federal resources may be necessary to mitigate a chemical/biological incident, local public safety agencies will be the first on the scene. Advanced plans must be made for coordinated interagency and intergovernmental cooperation during response.**

In this incident, civilian resources were effectively utilized to mitigate the incident. Military advisors were contacted for assistance, and military facilities were used to test the products. The FBI assumed its role without alienating fire/EMS personnel. The need for interagency coordination and cooperation is paramount, but it also must be remembered that the first responders will be uniformed civilian personnel: police, fire, and EMS. In most parts of the country, Federal resources will be deployed from distant locations requiring several hours to become operational.

9. **Standard operating procedures are necessary for response to chemical and biological terrorism.**

The DCFEMS was able to successfully adapt its existing Hazmat SOPs and a terrorism check-list into a working action plan for this incident. All public safety departments need to develop SOPs or other operational guidelines for response to chemical and biological threats. These SOPs should identify key initial operations to maintain the safety of responders and minimize the scope and impact of an incident, as well as list available local, State and Federal resources and contacts that may be useful during operations. A "response to terrorism" SOP should incorporate critical aspects of Hazmat, mass casualty, command, and emergency management operations. DCFEMS has included a check-list for response to terrorist incidents, including explosives and NBC weapons, into its incident command system. Fire commanders have trained for bomb and NBC response since 1996.

10. Hazmat teams need special equipment to respond to these types of incidents.

Due to budget constraints beyond the control of DCFEMS personnel, Hazmat crews were hampered by some equipment shortages, notably tents to protect the privacy of people being decontaminated. Additional trained personnel, protective equipment, supplied-air breathing apparatus, and communications gear would have aided Hazmat crews.

11. Good public information is key for helping to maintain public confidence in emergency personnel and emergency organizations.

The terrorists' goal is to make people feel unsafe and unprotected. Successful public information management utilizes the media to relay important information to the public as well as to help solidify the public's confidence in the responding agencies. A technically competent person from the special operations groups involved should be assigned to explain technical operations in plain terms to the media.

The fire department PIO at such an incident needs to be supported with a technical expert to handle media inquiries in an appropriate fashion. Because of the lack of a technical spokesperson, the media turned to experts not directly involved for commentary. This practice encouraged people to speculate widely on what was occurring at this incident, and resulted in incorrect and conflicting information to be transmitted to the public.

12. Fire and EMS agencies should anticipate the requirements associated with quarantining a sizeable portion of the exposed population after a biological incident.

A Biological weapon release may necessitate isolating any persons suspected of exposure for up to several days. The logistical and legal ramifications of such a task need to be reviewed. The DCFEMS isolated the small number of exposed persons at this incident. Departments should consider adding equipment such as large tents to their resource lists for temporary sheltering of victims.

13. A direct phone line should be established and kept open with those who are protected in place. Information should be channeled through a single source from the incident management team.

The DCFEMS established communications with the people inside the B'nai B'rith building, but incident managers indicated that this line of communication was not kept open. This enabled conflicting information to be communicated to the people inside by other responding agencies. At any incident of this type, a single person or agency should be established as the contact for victims to ensure the delivery of consistent information and instructions from the incident managers.

14. **Fire and EMS agencies should maintain an operational command post separate from the unified command post.**

DCFEMS commanders indicated that they would establish a separate operational command post away from the unified command post on future incidents to facilitate fire/EMS operations. The command post at this incident was crowded and made smooth organization of the actual fire service operations difficult. Fire departments should have plans for their own command post area so they can effectively coordinate the strategies that are developed in the unified command post. Additionally, access to the unified command post should be limited to key incident management team members and liaisons from responding agencies. Security is necessary at the fire and unified command posts to restrict access.

15. **Thorough documentation and record keeping should take place concurrently and after such a incident.**

Diverse information was collected for this report. Some information was contradictory or inaccurate. Other information, such as names of personnel, the decisions behind action plans and mitigation strategies, existed only in the memories of the responders. It is important and valuable for departments to develop a process to document all the decisions and actions made by personnel during such incidents. This documentation should preferably begin during the response to the incident, and continue after the incident as part of a comprehensive post-incident analysis. This would allow departments to develop plans for improving operations, provide a legal record of what occurred, and preserve as part of the institutional memory the lessons learned by the department, long after the current responders have left service.

The Federal Emergency Management Agency's Urban Search and Rescue system has developed criteria for Technical Documentation personnel that could provide an excellent model for incorporating documentation of incidents into local department's response plans.

16. **First responders need affordable equipment to detect and characterize chemical and biological agents.**

Devices and field test kits that are currently being developed for the military and Federal government should be made available to civilian first responders. Currently affordable technology is limited and some types are prone to false positive readings. More funding should be provided to develop promising technologies. The ability to detect the release of chemical and biological agents was recently identified as a priority in a survey of needs of State and local responders by the US Department of Justice.[2]

17. **More research is necessary to determine the most appropriate strategies for protecting the public during these incidents.**

More research is needed to address the question of whether it is best to protect in-place or to evacuate to a safe haven, those civilians exposed to chem/bio agents.

---

[2] Inventory of State and Local Enforcement Technology Needs to Combat Terrorism, US Department of Justice, 1998.

INITIATIVES BY DCFEMS

The DCFEMS has taken several initiatives to improve response in the future to these incidents:

- Formed a Special Operations section and added a battalion chief for Special Operations to oversee technical rescue, Hazmat, and counter terrorism;

- Purchased additional level A encapsulated suits;

- Purchased chemical/biological filter masks;

- Trained all personnel in the First Battalion as Hazmat Technicians for suit operations;

- Added the Council of Government's MMST to response plans;

- Purchased decontamination tents to protect the privacy of victims;

- Providing first response to terrorism training for commanders and firefighters;

- Purchasing new communications equipment for the Hazmat team;

- Reviewing and revising all incident command systems and SOPs, including the terrorism response checklist.

District of Columbia Fire and EMS Department Incident Command System and Checklist

INCIDENT COMMAND SYSTEM FOR TERRORIST ATTACKS

Terrorism is a new challenge to the emergency services in this country that will tax every segment of the public safety community. The size of these incidents, the number of agencies involved, the need to preserve a very large crime scene, the number of casualties and entrapped victims, the possibility of contamination of the entire community, and injury to emergency personnel from contaminants or secondary explosions are problems unique to terrorist incidents. Pre-planning and cooperation between Federal, State, and local authorities will be essential to successfully mitigate this type emergency. The Federal Emergency Management Agency (FEMA) held a conference in November 1995 to address the Nation's preparedness to deal with terrorist activity. At the conclusion of the conference, Strengthening the Fire and Emergency Response to Terrorism, it was resolved that Nationwide focus must be directed at the terrorist threat so communities will be properly prepared.

> There was a broad scope of issues presented at the conference with presenters from France, Israel, Japan, Ireland, and England representing the international community. Federal agencies including the Federal Bureau of Investigation, State Department, Department of Defense, Department of Energy, U.S. Fire Administration, Nuclear Emergency Management Laboratories, U.S. Army, U.S. Marines, and the Office of Emergency Preparedness were represented. Chief officers from fire departments representing the entire Nation (including Oklahoma City and New York City which experienced terrorist attacks), professional associations such as the International Association of Fire Chiefs and the International Association of Chiefs of Police, and prominent figures from the media were part of the roster. Virtually all aspects of the terrorist issue were addressed from as many perspectives as possible. The consensus at the conference was that this was just the beginning of a Nationwide effort to prepare to meet the challenge.

There was one common chorus that sounded throughout the conference. A comprehensive, well organized, and efficiently implemented Incident Command System would be of paramount importance to successfully manage any large scale emergency incident, particularly one dealing with multiple agencies. It was also the consensus that the Fire Service had the most sophisticated Incident Command System and the most experience in applying it. For this reason fire departments across the Nation will be called upon to lead the way in mitigating incidents resulting from acts of terrorism. The check-off list below should be helpful in applying any Incident Command System to a terrorist attack.

15

Some unique problems to be considered by the incident commander in applying the Incident Command System to a terrorist attack would include:

1. Secondary explosions while rescuers are in the area
2. The presence of biochemical contaminants
3. The presence of radiological contaminants
4. Preserving the scene for criminal investigation
5. Coordination of multi agency response
6. Resources for a long duration operation
7. Controlling access to restricted areas
8. Security of adjacent buildings and property
9. Notification of relatives
10. Media briefings for a large scale media response

The check lists below are intended to remind the incident commander and sector leaders of some considerations that may be unique to a terrorist attack.

PROBLEM AREA	SECTOR ASSIGNMENT	CONSIDERATIONS
Limit Access	Operations	____ Minimal Commitment Of Personnel
Secondary Explosion	Operations	____ 1. Bomb Disposal Unit
		____ 2. Building Search
		____ 3. Snatch Rescue
		____ 4. Specialized Unit
Contaminants	Operations	____ 1. Hazmat
		____ 2. Chemical
		____ 3. Biological
		____ 4. Radiological
Search & Rescue	Operations	____ 1. Snatch Rescues
		____ 2. Primary
		____ 3. Secondary
EMS	Medical	____ 1. Triage
		____ 2. Treatment
		____ 3. Overloading Facilities
		____ 4. Transport
		____ 5. Record Keeping

PROBLEM AREA	SECTOR ASSIGNMENT	CONSIDERATIONS
Evacuation	Planning	____ 1. On-site Treatment
		____ 2. Contamination of Facilities
		____ 3. Safe Zones
		____ 4. Child Care
		____ 5. Mass Transit
		____ 6. Record Keeping
Perimeters	Planning	____ 1. Primary
		____ 2. Secondary
		____ 3. Large Enough for Specialty Equip.
Zones	Planning	____ 1. Hot
		____ 2. Warm
		____ 3. Cold
		____ 4. Notification of All Agencies
		____ 5. Record Keeping
Other Agencies	Planning	____ 1. Mutual Aid
		____ 2. Federal
		____ 3. State
		____ 4. Military
		____ 5. Statutory Auth.
Communication	Communication	____ 1. Special Phones
		____ 2. Compatibility
		____ 3. Messenger
		____ 4. Command Post Liaison
Foam & Other Agents	Operation	____ 1. Type
		____ 2. Compatibility
		____ 3. Effectiveness
		____ 4. Complications
Logistics	Logistics	____ 1. Food On Site/Off Site
		____ 2. Fuel
		____ 3. Accountability
		____ 4. Storage
		____ 5. Santitation
		____ 6. Water
		____ 7. Waste

continued on next page

PROBLEM AREA	SECTOR ASSIGNMENT	CONSIDERATIONS
Logistics *(continued)*	Logistics	____ 8. Contamination
		____ 9. Clothing
		____ 10. Donations
		____ 11. Purchases
		____ 12. Loans
		____ 13. Special Equip.
		____ 14. Repair
		____ 15. Parts
		____ 16. Replacement
		____ 17. Contracts
		____ 18. Record Keeping
		____ 19. Delivery Zone
Public Information	P.I.O.	____ 1. Spokesperson
		____ 2. Location
		____ 3. Access
		____ 4. Other Agencies
		____ 5. Time
Special Teams	Planning	____ 1. Specialty
		____ 2. Command
		____ 3. Equipment
		____ 4. Response Time
		____ 5. Communications
		____ 6. Relief
Morgue	Medical	____ 1. Criminal Invest.
		____ 2. Record Keeping
		____ 3. Restricted Access
		____ 4. Temp/weather
Briefing	P.I.O.	____ 1. Personnel
		____ 2. Other Agencies
		____ 3. Survivors
Relief	Planning	____ 1. Short Term
		____ 2. Long Term (8 Dys)
		____ 3. Expertise

PROBLEM AREA	SECTOR ASSIGNMENT	CONSIDERATIONS
Family Assistance	Liaison	____ 1. Survivors ____ 2. Relatives ____ 3. Briefing ____ 4. Debriefing ____ 5. Transportation ____ 6. Lodging ____ 7. Subsistence
Finance	Finance	____ 1. Authority To Request ____ 2. Record Keeping ____ 3. Accountability ____ 4. Payment ____ 5. Cost ____ 6. Claims ____ 7. Analysis ____ 8. Other Agencies
Debriefing	P.I.O.	____ 1. Right To Know ____ 2. Personnel ____ 3. Survivors
Physical Therapy	Liaison	____ 1. Rescue Workers ____ 2. On Site ____ 3. Off Site
C.I.S.D.	Liaison	____ 1. Critical Incident Stress Debriefing ____ 2. Additional Teams ____ 3. Follow Up
Counselling	Liaison	____ 1. Personnel ____ 2. Relatives ____ 3. Survivors ____ 4. Next Of Kin ____ 5. Witnesses

APPENDIX B

B'nai B'rith Incident Pictures

(Photos Courtesy DCFEMS)

Photo 1. Firefighters establish a decontamination corridor

Photo 2. Hazardous materials crews prepare decontamination area.

Appendix B (continued)

Photo 3. Command post operations were crowded by responders from multiple agencies.

Photo 4. The decon area is set up, consisting of a tool drop station, gross bleach wash, rinse, and a second wash and rinse area.

Appendix B (continued)

Photo 5. A civilian undergoes decontamination.

Appendix B (continued)

Photo 6. A patient is transported suffering from chest pains during the incident.

Photo 7. An overview of apparatus placement. The arrow points to the
B'nai B'rith Building.

APPENDIX C

HANDBOOK ON THE MEDICAL ASPECTS OF NBC DEFENSIVE OPERATIONS FM 8-9

PART II – BIOLOGICAL

ANNEX B

CLINICAL DATA SHEETS FOR SELECTED BIOLOGICAL AGENTS

1 February 1996

B.02 Anthrax

B.09 Plague

B.02 ANTHRAX

a.   Clinical Syndrome.

(1) Characteristics. Anthrax is a zoonotic disease caused by Bacillus anthracis. Under natural conditions, humans become infected by contact with infected animals or contaminated animal products. Human anthrax is usually manifested by cutaneous lesions. A biological warfare attack with anthrax spores delivered by aerosol would cause inhalation anthrax, an extraordinary rare form of the naturally occurring disease.

(2) Clinical Features. The disease begins after an incubation period varying from 1-6 days, presumably dependent upon the dose of inhaled organisms. Onset is gradual and nonspecific, with fever, malaise, and fatigue, sometimes in association with a nonproductive cough and mild chest discomfort. In some cases, there may be a short period of improvement. The initial symptoms are followed by in 2-3 days by the abrupt development of severe respiratory distress with dyspnea, diaphoresis, stridor, and cyanosis. Physical findings may include evidence of pleural effusions, edema of the chest wall, and meningitis. Chest x-ray reveals a dramatically widened mediastinum, often with pleural effusions, but typically without infiltrates. Shock and death usually follow within 24-36 hours of respiratory distress onset.

b.   Diagnosis

(1) Routine Laboratory Findings. Laboratory evaluation will reveal a neutrophilic leucocytosis. Pleural and cerebrospinal fluids may be hemorrhagic.

(2) Differential Diagnosis. An epidemic of inhalation anthrax in its early state with nonspecific symptoms could be confused with a wide variety of viral, bacterial, and fungal infections. Progression over 2-3 days with the sudden development of severe respiratory distress followed by shock and death in 24-36 hours is essentially all untreated cases eliminates diagnoses other than inhalation anthrax. The presence of a widened mediastinum on chest x-ray, in particular, should alert one to the diagnosis. Other suggestive findings include chest-wall edema, hemorrhagic pleural effusions, and hemorrhagic meningitis. Other diagnoses to consider include aerosol exposure to SEB; but in this case onset would be more rapid after exposure (if known), and no prodrome would be evident prior to onset of sever respiratory symptoms. Mediastinal widening on chest x-ray will also be absent. Patients with plague or tularemia pneumonia will have pulmonary infiltrates and clinical signs of pneumonia (usually absent in anthrax).

(3) Specific Laboratory Diagnosis. Bacillus anthracis will be readily detectable by blood culture with routine media. Smears and cultures of pleural fluid and abnormal cerebrospinal fluid may also be positive. Impression smears of mediastinal lymph nodes and spleen from fatal cases should be positive. Toxemia is sufficient to permit anthrax toxin detection in blood by immunoassays.

c.   Therapy. Almost all cases of inhalation anthrax in which treatment was begun after patients were symptomatic have been fatal, regardless of treatment. Historically, penicillin has been regarded as the treatment of choice, with 2 million units given intravenously every 2 hours. Tetracyclines and erythromycin have been recommended in penicillin-sensitive patients. The vast majority of anthrax strains are sensitive in vitro to penicillin. However, penicillin-resistant strains exist naturally, and one has been recovered from a fatal human case. Moreover, it is not difficult to induce resistance to penicillin, tetracyclines, erythromycin, and many other antibiotics through laboratory manipulation of organisms. All naturally occurring strains tested to date have been sensitive to erythromycin, chloramphenicol, gentamicin, and ciprofloxacin. In the absence of information concerning antibiotic sensitivity, treatment should be instituted at the earliest signs of disease with oral ciprofloxacin (1,000 mg initially, followed by 750 mg po (orally) bid (twice daily) or intravenous doxycycline (200 mg initially, followed by 100 mg q (every 12 hrs.) Supportive therapy for shock, fluid volume deficit, and adequacy of airway may all be needed.

d. Prophylaxis

(1) Vaccine. A licensed, alum-precipitated preparation of purified B. anthracis protective antigen (PA) has been shown to be effective in preventing or significantly reducing the incidence of inhalation anthrax. Limited human data suggest that after completion of the first three doses of the recommended six-dose primary series (0, 2, 4, weeks, then 6, 12, 18 months), protection against both cutaneous and inhalation anthrax is afforded. Studies in rhesus monkeys indicate that good protection is afforded after two doses (10-16 days apart) for up to 2 years. It is likely that two doses in humans is protective as well, but there is too little information to draw firm conclusions. As will all vaccines, the degree of protection depends upon the magnitude of the challenge dose; vaccine-induced protection is undoubtedly overwhelmed by extremely high spore challenge. At least three doses of the vaccine (at 0, 2, and 4 weeks) are recommended for prophylaxis against inhalation anthrax. Contraindications for use are sensitivity to vaccine components (formalin, alum, benzethonium chloride) and/or history of clinical anthrax. Reactogenicity is mild to moderated: up to 6% of recipients will experience mild discomfort at the inoculation site for up to 72 hours (tenderness, erythema, edeman, pruritus), while a smaller proportion (<1%) will experience more severe local reactions (potentially limiting use of the extremity for 1-2 days); modest systemic reactions (anaphylaxis, which precludes additional vaccination) are rare. The vaccine should be stored at refrigerator temperature (not frozen).

(2) Antibiotics. Choice of antibiotics for prophylaxis is guided by the same principle as that for treatment; i.e., it is relatively easy to produce a penicillin-resistant organism in the laboratory, and possible, albeit somewhat more difficult, to induce tetracycline resistance. Therefore, if there is information indicating that a biological weapon attack is imminent, prophylaxis with ciprofloxacin (500 mg po bid), or doxycycline (100 mg po bid) is recommended. If unvaccinated, a single 0.5 ml dose of vaccine should also be given subcutaneously. Should the attack be confirmed as anthrax, antibiotics should be continued for at least 4 weeks in all exposed. In addition, two 0.5 ml doses of vaccine should be given 2 weeks apart in the unvaccinated; those previously vaccinated with fewer than three doses should receive a single 0.5 ml booster, while vaccination probably is not necessary for those who have received the initial three doses within the previous 6 months (primary series). Upon discontinuation of antibiotics, patients should be closely observed; if clinical signs of anthrax occur, patients should be treated as indicated above. If vaccine is not available, antibiotics should be continued beyond 4 weeks until the patient can be closely observed upon discontinuation of therapy.

B.09 PLAGUE

a. Clinical Syndrome

(1) Characteristics. Plague is a zoonotic disease caused by Yersinia pestis. Under natural conditions, humans become infected as a result of contact with rodents, and their fleas. The transmission of the gram-negative coccobacillus is by the bite of the infected flea, Xenopsylla cheopis, the oriental rat flea, or Pulex irritans, the human flea. Under natural conditions, three syndromes are recognized: bubonic, primary septicemic, or pneumonic. In a biological warfare scenario, the plague bacillus could be delivered via contaminated vectors (fleas) causing the bubonic type or, more likely, via aerosol causing pneumonic type.

(2) Clinical Features. In bubonic plague, the incubation period ranges from 2 to 10 days. The onset is acute and often fulminant with malaise, high fever, and one or more tender lymph nodes. Inguinal lymphadenitis (bubo) predominates, but cervical and axillary lymph nodes can progress spontaneously to the septicemic form with organisms spread to the central nervous system, lungs (producing pneumonic disease), and elsewhere. The mortality is 50 percent in untreated patients with the terminal event being circulatory collapse, hemorrhage, and peripheral thrombosis. In primary pneumonic plague, the incubation period is 2 to 3 days. The onset is acute and fulminant with malaise, high fever, chills, headache, myalgia, cough with production of bloody sputum, and toxemia. The pneumonia progresses rapidly, resulting in dyspnea, stridor, and cyanosis. In untreated patients, the mortality is 100 percent with the terminal event being respiratory failure, circulatory collapse, and a bleeding diathesis.

b. Diagnosis

(1) Presumptive. Presumptive diagnosis can be made by identification of the gram-negative coccobacillus with safety-pin bipolar staining organisms in Giemsa or Wayson's stained slides from a lymph node needle aspirate, sputum, or cerebrospinal fluid (CSF) samples. When available, immunofluorescent staining is very useful. Elevated levels of antibody to Y. pestis in a nonvaccinated patient may also be useful.

(2) Definitive. Yersina pestis can be readily cultured from blood, sputum, and bubo aspirates. Most naturally occurring strains of Y. pestis produce an "F1" antigen in vivo which can be detected in serum samples by immunoassay. A fourfold rise of Y. pestis antibody levels in patient serum is also diagnostic.

(3) Differential. In cases where bubonic type is suspected, tularemia adentitis, staphylococcal or streptococcal adentitis, meningococcemai, enteric gram-negative sepsis, and rickettsiosis need to be ruled out. In pneumonic plague, tularemia, anthrax, and staphylococcal enterotoxin B (SEB) agents need to be considered. Continued deterioration without stabilization effectively rules out SEB. The presence of a widened meiastinum on chest x-ray should alert one to the diagnosis of anthrax.

c. Therapy. Plague may be spread from person to person by droplets. Strict isolation procedures for all cases are indicted. Streptomycin, tetracycline, and chloramphenicol are highly effective if begun early. Significant reduction in morbidity and mortality is possible if antibiotics are given within the first 24 hours after symptoms of pneumonic plague develop. Intravenous doxycycline (200 mg initially, followed by 100 mg every 12 hours), intramuscular streptomycin (1 g every 12 hours), or intravenous chloramphenicol (1 g every 6 hours) for 10-14 days are effective against naturally occurring strains. Supportive management of life-threatening complications from the infection, such as shock, hyperpyrexia, convulsions, and disseminated intravascular coagulation (DIC), need to be initiated as they develop.

d.   *Prophylaxis.* A formalin-killed Y. *pestis* vaccine is produced in the United States and has been extensively used. Efficacy against flea-borne plague is inferred from population studies, but the utility of this vaccine against aerosol challenge is unknown. Reactogenicity is moderately high and a measurable immune response is usually attained after a 3-dose primary series: at 0, 1, and 4-7 months. To maintain immunity, boosters every 1-2 years are required. Live-attenuated vaccines are available elsewhere but are highly reactogenic and without proven efficacy against aerosol challenge.

Incident Operations Chart

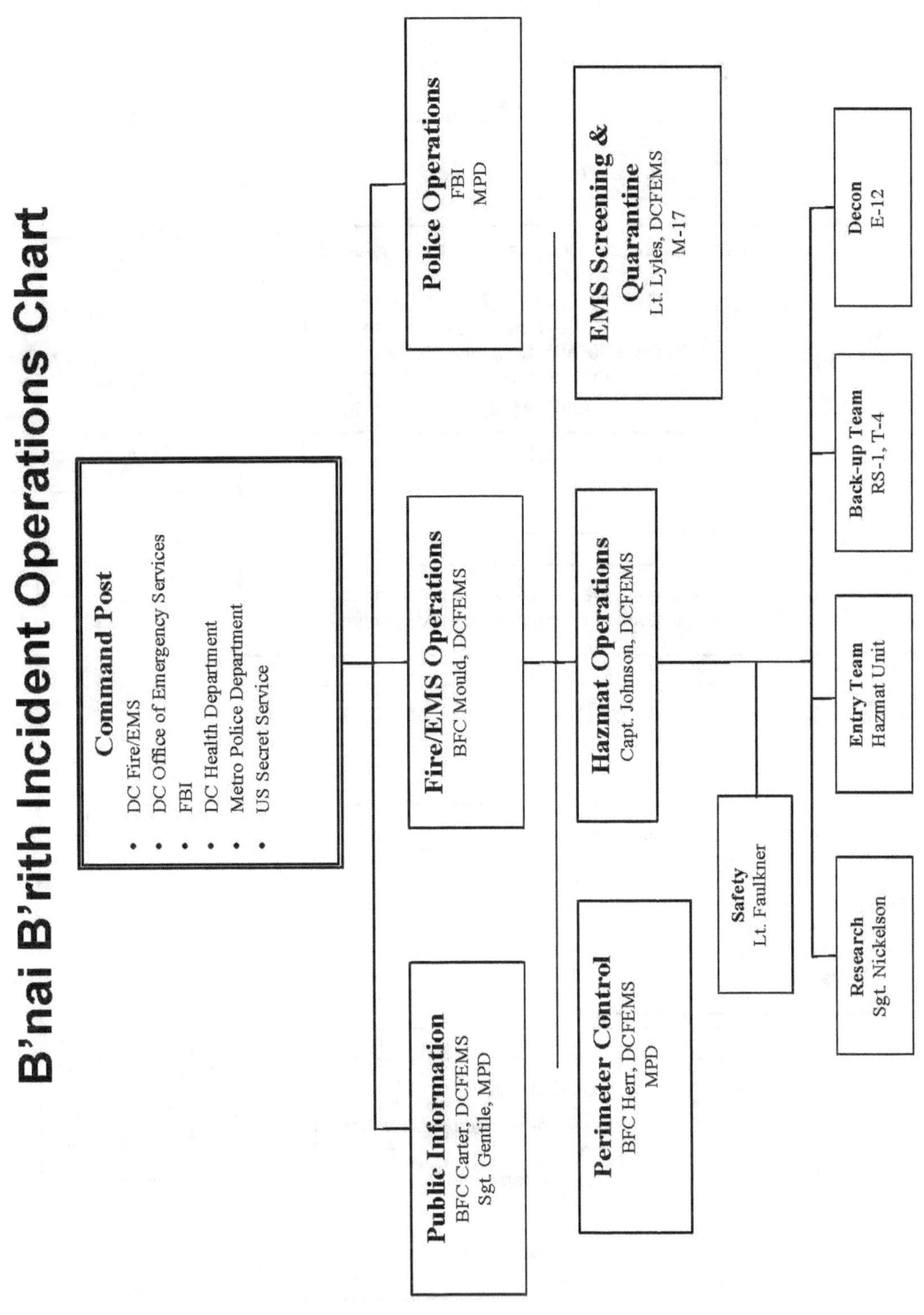

B'nai B'rith Incident Operations Chart

Command Post
DC Fire/EMS
DC Office of Emergency Services
FBI
DC Health Department
Metro Police Department
US Secret Service

Police Operations
FBI
MPD

Fire/EMS Operations
BFC Mould, DCFEMS

Public Information
BFC Carter, DCFEMS
Sgt. Gentile, MPD

EMS Screening & Quarantine
Lt. Lyles, DCFEMS
M-17

Hazmat Operations
Capt. Johnson, DCFEMS

Perimeter Control
BFC Herr, DCFEMS
MPD

Safety
Lt. Faulkner

Entry Team
Hazmat Unit

Back-up Team
RS-1, T-4

Decon
E-12

Research
Sgt. Nickelson

APPENDIX E

Model Action Plan Chart for Chemical/Biological Incidents for First Responders

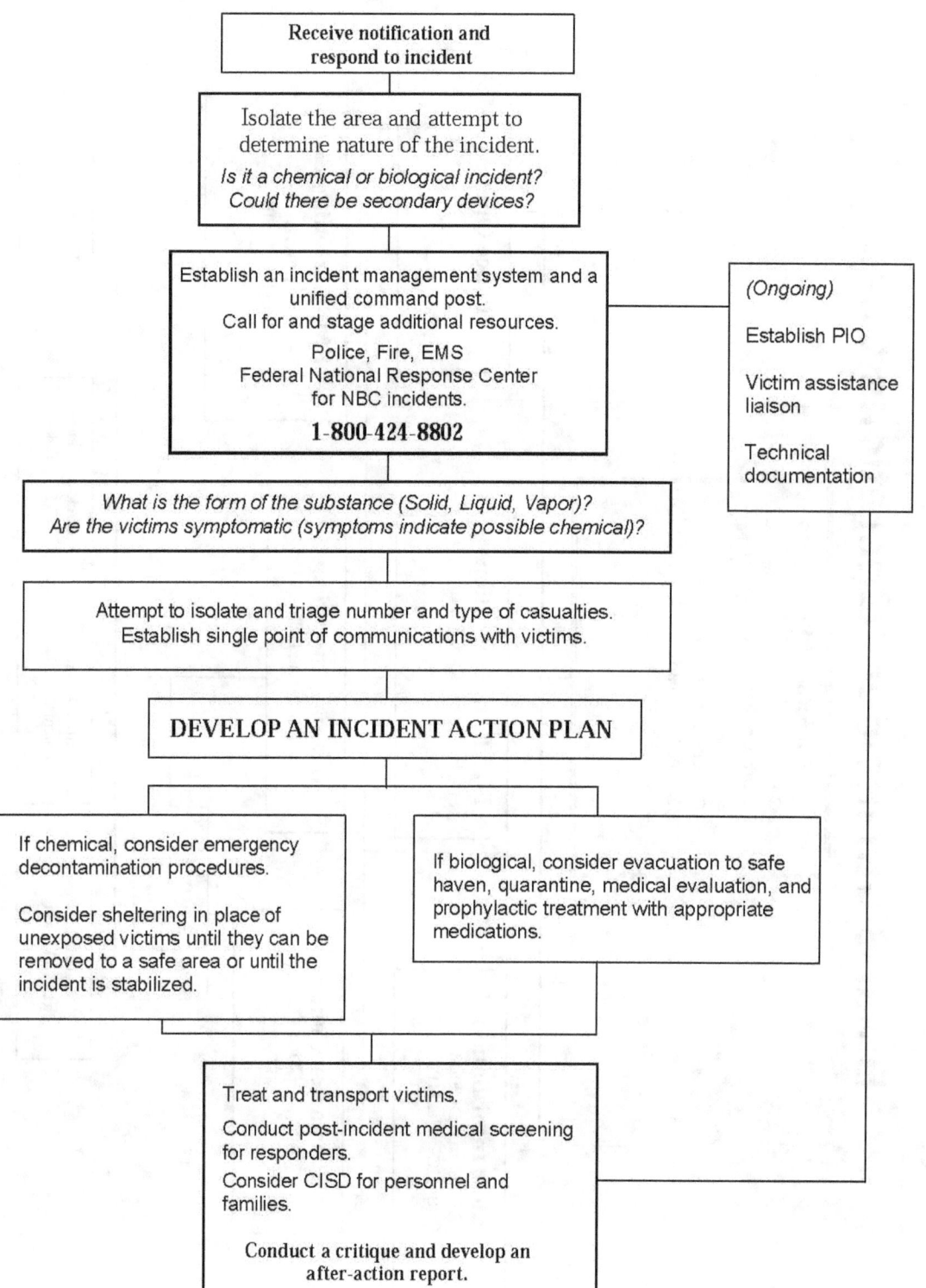

Receive notification and respond to incident

Isolate the area and attempt to determine nature of the incident.
Is it a chemical or biological incident? Could there be secondary devices?

Establish an incident management system and a unified command post.
Call for and stage additional resources.
Police, Fire, EMS
Federal National Response Center for NBC incidents.
1-800-424-8802

(Ongoing)

Establish PIO

Victim assistance liaison

Technical documentation

What is the form of the substance (Solid, Liquid, Vapor)? Are the victims symptomatic (symptoms indicate possible chemical)?

Attempt to isolate and triage number and type of casualties.
Establish single point of communications with victims.

DEVELOP AN INCIDENT ACTION PLAN

If chemical, consider emergency decontamination procedures.

Consider sheltering in place of unexposed victims until they can be removed to a safe area or until the incident is stabilized.

If biological, consider evacuation to safe haven, quarantine, medical evaluation, and prophylactic treatment with appropriate medications.

Treat and transport victims.

Conduct post-incident medical screening for responders.

Consider CISD for personnel and families.

Conduct a critique and develop an after-action report.